Energy Sector Standard of the People's Republic of China

NB/T 31112-2017

Technical requirements for tender design of wind power projects

风电场工程招标设计技术规定

(English Translation)

China Water & Power Press

Beijing 2024

All rights reserved. No part of this publication may be reproduced, stored in a retrieval system, or transmitted in any form or by any means—electronic, mechanical, photocopying, recording or otherwise, without prior written permission of the publisher.

图书在版编目（CIP）数据

风电场工程招标设计技术规定 = Technical requirements for tender design of wind power projects（NB/T 31112-2017）：英文 / 国家能源局发布. -- 北京：中国水利水电出版社，2024.6. -- ISBN 978-7-5226-2591-1

I. TM614-65

中国国家版本馆CIP数据核字第2024Y3Z294号

Energy Sector Standard of the People's Republic of China

中华人民共和国能源行业标准

Technical requirements for tender design of wind power projects

风电场工程招标设计技术规定

NB/T 31112-2017

(English Translation)

Issued by National Energy Administration of the People's Republic of China
国家能源局　发布
Translation organized by China Renewable Energy Engineering Institute
水电水利规划设计总院　组织翻译
Published by China Water & Power Press
中国水利水电出版社　出版发行
　　Tel: (+ 86 10) 68545888　68545874
　　sales@mwr.gov.cn
　　Account name: China Water & Power Press
　　Address: No.1, Yuyuantan Nanlu, Haidian District, Beijing 100038, China
　　http: //www.waterpub.com.cn
中国水利水电出版社微机排版中心　排版
北京中献拓方科技发展有限公司　印刷
210mm×297mm　16开本　1.75印张　70千字
2024年6月第1版　2024年6月第1次印刷
Price(定价)：￥280.00

About English Translation

This English version is one of China's energy sector standard series in English. Its translation was organized by China Renewable Energy Engineering Institute authorized by National Energy Administration of the People's Republic of China in compliance with relevant procedures and stipulations. This English version was issued by National Energy Administration of the People's Republic of China in Announcement [2021] No. 5 dated November 16, 2021.

This version was translated from the Chinese Standard NB/T 31112-2017, *Technical requirements for tender design of wind power projects*, published by China Water & Power Press. The copyright is reserved by National Energy Administration of the People's Republic of China. In the event of any discrepancy in the implementation, the Chinese version shall prevail.

Many thanks go to the staff from relevant standard development organizations and those who have provided generous assistance in the translation and review process.

For further improvement of the English version, all comments and suggestions are welcome and should be addressed to:

China Renewable Energy Engineering Institute
No. 2 Beixiaojie, Liupukang, Xicheng District, Beijing 100120, China
Website: www.creei.cn

Translating organization:

POWERCHINA Zhongnan Engineering Corporation Limited

Translating staff:

CHEN Lei   YANG Hong   LI Qian

Review panel members:

QIE Chunsheng	Senior English Translator
JIN Feng	Tsinghua University
YAN Wenjun	Army Academy of Armored Forces, PLA
GUO Jie	POWERCHINA Beijing Engineering Corporation Limited
LI Zhongjie	POWERCHINA Northwest Engineering Corporation Limited
LI Yu	POWERCHINA Huadong Engineering Corporation Limited
LIU Xiaofen	POWERCHINA Zhongnan Engineering Corporation Limited
MI Youwan	POWERCHINA Zhongnan Engineering Corporation Limited

National Energy Administration of the People's Republic of China

翻译出版说明

本译本为国家能源局委托水电水利规划设计总院按照有关程序和规定，统一组织翻译的能源行业标准英文版系列译本之一。2021年11月16日，国家能源局以2021年第5号公告予以公布。

本译本是根据中国水利水电出版社出版的《风电场工程招标设计技术规定》NB/T 31112—2017翻译的，著作权归国家能源局所有。在使用过程中，如出现异议，以中文版为准。

本译本在翻译和审核过程中，本标准编制单位及编制组有关成员给予了积极协助。

为不断提高本译本的质量，欢迎使用者提出意见和建议，并反馈给水电水利规划设计总院。

地址：北京市西城区六铺炕北小街2号
邮编：100120
网址：www.creei.cn

本译本翻译单位：中国电建集团中南勘测设计研究院有限公司

本译本翻译人员：陈　蕾　杨　虹　李　倩

本译本审核人员：

郄春生　英语高级翻译

金　峰　清华大学

闫文军　中国人民解放军陆军装甲兵学院

郭　洁　中国电建集团北京勘测设计研究院有限公司

李仲杰　中国电建集团西北勘测设计研究院有限公司

李　瑜　中国电建集团华东勘测设计研究院有限公司

刘小芬　中国电建集团中南勘测设计研究院有限公司

糜又晚　中国电建集团中南勘测设计研究院有限公司

国家能源局

Contents

Foreword		V
1	Scope	1
2	Normative references	1
3	Terms and definitions	1
4	General provisions	2
5	Engineering investigation	2
5.1	Engineering survey	2
5.2	Engineering geological investigation	2
5.3	Hydrology	3
6	Micrositing	3
6.1	Wind resource evaluation	3
6.2	Wind turbine layout	3
6.3	Verification of wind turbine layout	3
7	Lotting	4
7.1	Lotting principles	4
7.2	Lotting scheme	4
7.3	Sequence of tendering	5
8	Procurement and installation of wind turbine	5
8.1	Procurement of wind turbine proper	5
8.2	Procurement of tower	6
8.3	Installation of wind turbine	7
9	Foundations for wind turbine and its step-up equipment	8
9.1	Design scheme	8
9.2	Scope of tender	8
9.3	Main technical data	8
9.4	Main technical requirements	9
10	Procurement and installation of wind turbine step-up equipment	9
10.1	Procurement of wind turbine step-up equipment	9
10.2	Installation of wind turbine step-up equipment	10
11	Power collection system	10
11.1	Design scheme	10
11.2	Scope of tender	10
11.3	Main technical data	11
11.4	Main technical requirements	11
12	Road and wind turbine hardstand	11
12.1	Design scheme	11
12.2	Scope of tender	12
12.3	Main technical data	12
12.4	Main technical requirements	12
13	Step-up substation	13
13.1	Procurement of electrical equipment	13
13.2	Installation of electrical equipment	13
13.3	Civil works	14
13.4	Fire protection	16

Foreword

This standard is drafted in accordance with the rules given in the GB/T 1.1-2009 *Directives for standardization—Part 1: Structure and drafting of standards*.

National Energy Administration of the People's Republic of China is in charge of the administration of this standard. China Renewable Energy Engineering Institute has proposed this standard and is responsible for its routine management. Sub-committee on Planning and Design of Wind Power Project of Energy Sector Standardization Technical Committee on Wind Power is responsible for the interpretation of the specific technical content. Comments and suggestions in the implementation of this standard should be addressed to:

China Renewable Energy Engineering Institute
No. 2 Beixiaojie, Liupukang, Xicheng District, Beijing 100120, China

Chief development organizations:

POWERCHINA Zhongnan Engineering Corporation Limited

HydroChina Corporation

Chief drafting staff:

MI Youwan	ZHU Yijun	LIU Xiaosong	HUANG Chunfang
LIU Guopin	FU Liangming	YAN Biao	WU Jinhua
CHEN Min	WANG Dailan	CHEN Guibin	LIU Lihong
TAN Zhengguang	DU Jianwen	PENG Jiali	LIU Linqi
LUO Chengxi	LIU Changqing	SUN Jiansong	HE Xudong
ZHANG Qing	CAO Yuanyuan	WANG Sheng	

NB/T 31112-2017

Technical requirements for tender design of wind power projects

1 Scope

This standard specifies the workflow, lotting scheme, and main technical requirements for the tender design of wind power projects.

This standard is applicable to the tender design of grid-connected onshore wind power projects.

2 Normative references

The following referenced documents are indispensable for the application of this document. For dated references, only the edition cited applies. For undated references, the latest edition of the referenced document (including any amendments) applies.

GB 3096, *Environmental quality standard for noise*

GB/T 6067.1, *Safety rules for lifting appliances—Part 1: General*

GB/T 18451.1, *Wind turbine generator systems—Design requirements*

GB/T 18709, *Methodology of wind energy resource measurement for wind farm*

GB/T 18710, *Methodology of wind energy resource assessment for wind farm*

GB/T 19071.1, *Asynchronous generator for wind turbine generator systems—Part 1: Technical condition*

GB/T 19072, *Tower of wind turbine generator system*

GB/T 19568, *Wind turbines—Assembling and installation regulation*

GB/T 23479.1, *Wind turbine—Double-fed asynchronous generator—Part 1: Technical specification*

GB/T 25389.1, *Wind turbine low-speed permanent magnet synchronous generator—Part 1: Technical conditions*

GB 50065, *Code for design of ac electrical installations earthing*

GB 51096, *Code for design of wind farm*

NB 31089, *Code for design of fire protection for wind farms*

NB/T 31030, *Specifications for engineering geological investigation of wind power projects*

JGJ 276, *Technical code for safety of lifting in construction*

3 Terms and definitions

For the purposes of this document, the following terms and definitions apply.

3.1

tender technical document

document that specifies the detailed technical requirements for the tendering project, and constitutes an integral part of the tender documents and an important basis for concluding the contract

3.2
works

component of the wind power project, which has its own design documentation, construction conditions, and functionality, but cannot independently realize the production capacity or project benefits upon its completion

3.3
wind turbine

power generation equipment that converts the wind energy into electric energy, consisting of the wind turbine proper and the tower

4 General provisions

4.1 The tender design of a wind power project shall observe the relevant policies, laws and regulations of China, and be carried out according to the project site, scale and approved grid connection scheme on the basis of feasibility study report and approval documents, taking into account the requirements for project implementation and management.

4.2 The basic task of the tender design for a wind power project is to carry out the tender design and prepare the tender documents according to the requirements of project tendering procurement and project implementation and management. The tender design mainly includes the preliminary work and the specific design. The preliminary work consists of engineering investigation, micrositing and lotting; and the specific design shall involve the wind turbine, foundations for wind turbine and its step-up equipment, installation of wind turbine, wind turbine step-up equipment, power collection system, road and wind turbine hardstand, and step-up substation.

4.3 The tender documents should consist of project overview, main technical requirements, bill of quantities or list of equipment procurement, drawings, and attachments.

5 Engineering investigation

5.1 Engineering survey

The accuracy of topographic mapping within the wind farm shall be determined according to the topographic conditions, and the scale may be 1 : 500 to 1 : 2 000. For the topographic mapping for the step-up substation, the scale shall not be inferior to 1 : 500.

For the plane and section survey for overhead power collection system, the scale should be 1 : 200 to 1 : 500 longitudinal and be 1 : 2 000 to 1 : 5 000 transversal. In the case of long flyover, long span, or passing a crowded lot or an important crossover, a plane survey shall be conducted on the local area concerned, and the scale should be 1 : 2 000 to 1 : 5 000.

5.2 Engineering geological investigation

5.2.1 Identify and classify the geological hazards that might exist at the site of the wind farm or be induced by the wind turbines, step-up substation and other buildings (structures), assess the severity of the geological hazards and the site stability and suitability, and propose the engineering countermeasures accordingly.

5.2.2 Ascertain the engineering geological and hydrogeological conditions of the substrata for the wind turbines and step-up substation; and select the bearing stratum. Present such indexes as physical and mechanical parameters and electrical resistivity of rock and soil mass and corrosivity of soil and groundwater.

5.2.3 Ascertain and assess the engineering geological conditions for roads, power collection system, and spoil areas.

5.2.4 Propose the exploitation and utilization conditions for natural construction materials, if necessary.

5.2.5 The methodology and technical requirements for engineering geological investigation shall comply with NB/T 31030.

5.3 Hydrology

The flood control criteria for buildings (structures) in the wind farm shall be checked as per GB 51096.

6 Micrositing

6.1 Wind resource evaluation

6.1.1 Supplementary data shall be collected, including the relevant meteorological data from the reference observation stations near the site of a wind farm, wind data from the masts at and around the site, and topographic maps of the site and its periphery, and the site wind resource parameters shall be checked as per GB/T 18709 and GB/T 18710.

6.1.2 The wind condition parameters, such as wind direction, wind speed, wind power density and turbulence intensity at hub height, shall be checked.

6.1.3 The 50-year maximum wind speed, the average wind speed, characteristic turbulence intensity and maximum inflow angle for each wind turbine site shall be analyzed, and the wind turbine class shall be checked as per GB/T 18451.1.

6.2 Wind turbine layout

6.2.1 The wind turbine layout scheme shall be worked out through techno-economic comparison according to the selected wind turbine type, taking into account the wind conditions, topography, geology, transport, installation, environmental impacts, and nature of the land.

6.2.2 The wind turbines shall not occupy any prime farmland, shall avoid the core area and buffer zone of nature reserve, shall not affect cultural relics, military-sensitive area, air traffic control zone, first-class protected area of surface drinking water, or bird migration corridor, and shall not sit on any identified important mineral resources. The wind turbines shall occupy as less as possible cultivated land and avoid, whenever possible, the areas under special protection as approved by law by the government authorities at provincial level or higher.

6.2.3 The layout shall comply with the noise requirements specified in GB 3096.

6.2.4 The layout shall meet the requirements for safety distance from important surrounding facilities.

6.2.5 A certain number of wind turbine sites shall be reserved in the wind turbine layout scheme.

6.3 Verification of wind turbine layout

6.3.1 The wind turbine sites shall be checked one by one on the site and be adjusted according to the site-specific conditions and the basic layout principles.

6.3.2 After on-site check, the wind turbine layout scheme shall be subjected to safety check under various wind conditions by the wind turbine manufacturer, and a safety verification report shall be issued accordingly.

7 Lotting

7.1 Lotting principles

7.1.1 Lotting for a wind power project shall be carried out according to the project characteristics, construction duration, social resources, and project owner's requirements.

7.1.2 Lotting shall consider the project construction management requirements and be conducive to the quality control, progress control, cost control, and safety management.

7.1.3 Lotting shall consider the current construction technology and available construction machinery for wind power projects.

7.1.4 Lotting for a wind power project may be carried out by works and specialty, and the works may further be broken down by specialty or by work quantity. A project may be broken down into such works as wind turbine, wind turbine foundation, wind turbine step-up equipment, power collection system, road, and step-up substation. The step-up substation works includes civil works and procurement and installation of electrical equipment.

7.2 Lotting scheme

7.2.1 The procurement of wind turbine should be divided into two lots, one for the wind turbine proper and one for the tower; and the installation of wind turbine should be a separate lot. When the wind turbine is connected to its foundation with anchor bolts, the procurement of anchor bolts and accessories should be a separate lot or may be incorporated into the lot for the tower procurement.

7.2.2 The foundations for wind turbine and its step-up equipment should be one lot.

7.2.3 The procurement of wind turbine step-up equipment should be one lot.

7.2.4 The procurement and construction of full overhead power collection system should be one lot. The procurement and construction of full cable power collection system, including LV cables and accessories for wind turbine, should be divided into two separate lots. For mixed power collection system of both overhead lines and cables, the procurement of power cables and optical cables should be one lot, and the construction of the system should be a separate lot. The installation of wind turbine step-up equipment should be incorporated into the lot for the construction of power collection system.

7.2.5 The site access road should be one lot, and the on-site road and the hardstand should be another lot.

7.2.6 Lotting for procurement of electrical equipment for the step-up substation should be as follows:

- a) The main transformer should be one lot, and the station-service transformer and the earthing transformer should be a separate lot or may be incorporated into the lot for the main transformer.
- b) The 66 kV and above distribution equipment should be one lot.
- c) The 10 kV to 35 kV distribution equipment should be one lot.
- d) The station-service LV distribution equipment may be a separate lot or be incorporated into the lot for the 10 kV to 35 kV distribution equipment or the lot for the AC and DC power supply systems.
- e) The dynamic reactive power compensator should be one lot.

- f) The diesel generator set should be one lot.
- g) The HV and LV power cables for step-up substation should be incorporated into the lot for the cable power collection system or may be a separate lot.
- h) The control and protection system for step-up substation should be one lot.
- i) The AC and DC power supply systems should be one lot.
- j) The wind power prediction and control system should be one lot.
- k) The dispatching automation equipment and the gateway metering system should be one lot.
- l) The video monitoring and security system should be one lot.
- m) The communication equipment should be one lot.
- n) The control cables may be a separate lot or be incorporated into the lot for the procurement of cables of power collection system.

7.2.7 For the step-up substation, the civil works and the procurement and installation of electrical equipment, water supply and drainage equipment, and HVAC equipment for the buildings should be one lot, and the installation of electrical equipment for the substation should be another lot. The two lots may be merged into one.

7.2.8 The procurement of heating equipment for the step-up substation may be a separate lot or be incorporated into the lot for the civil works of the substation.

7.2.9 The decoration of the step-up substation and its landscaping may be two separate lots.

7.2.10 The fire protection system of the step-up substation should be one lot.

7.2.11 The soil and water conservation and environmental protection works should be incorporated into the corresponding lots for the road and civil works, or may be a separate lot.

7.2.12 The lightning protection and earthing and monitoring may be a separate lot or be incorporated into other lot(s).

7.3 Sequence of tendering

After the procurement tender for wind turbine proper has been completed, the tendering for other lots is carried out according to the project construction schedule.

8 Procurement and installation of wind turbine

8.1 Procurement of wind turbine proper

8.1.1 Scope of tender

The scope of tender for procurement of wind turbine proper covers:

- a) Scope of supply.
- b) Technical data and services to be provided by the wind turbine manufacturer.

8.1.2 Main technical data

The tender documents shall provide the data on wind, environment, topography, geology and transport conditions as follows:

- a) Wind conditions, including the average wind speed of the representative year, 50-year maximum wind speed and turbulence intensity at hub height.

b) Environmental conditions, including the altitude, temperature, humidity, precipitation, air density, blowing sand, thunderstorm, salt mist, hail, icing, snow, solar radiation, and earthquakes.

c) Transport conditions, including the geographical location of the wind farm, site accessibility, means of transport, and design data for site access road and on-site road.

8.1.3 Main technical requirements

The tender documents shall specify the following technical requirements:

a) Range of wind turbine capacity, scope of supply, means of transport and handling, terms and place of delivery, technical services, and explicit interface between wind turbine and other equipment.

b) Scope of wind turbine supply, which should include blade, generator, hub (including pitch mechanism), main shaft, main bearing, coupler, gearbox, yaw system, hydraulic system, frequency converter, pitch system, brake system, cooling system, generator circuit breaker, main control cabinet, wind turbine SCADA system, spare parts, consumables, special tools for installation, and wind turbine auxiliaries. The wind turbine auxiliaries should include lightning protection devices, fasteners such as bolts for connection between nacelle and tower, between tower sections and between tower and foundation connector, control system, power cabinet, fire alarming and extinguishing system, data acquisition system for wind speed and direction, as well as wires, cables, conductor rails, optical cables, pigtails and pigtail boxes in the wind turbine proper and the tower, first aid kit, emergency lights and escape devices in the nacelle, and lifting devices such as climbing aids, service lift or auto climbing system, safety apparels and fall arresters in the tower.

c) Procurement of wind turbine, which shall cover technical data and services.

d) Interface between lots for wind turbine and electrical equipment.

e) Certifications and test certificates required by the competent national authorities.

f) Grid connection.

g) Selection of wind turbine and its technical parameters, which shall comply with GB/T 19071.1, GB/T 23479.1, GB/T 25389.1, and GB/T 25389.2.

h) Design service life of wind turbine.

i) Guaranteed static/dynamic power curve of each wind turbine, average availability of wind turbines and availability of single wind turbine in the warranty period.

8.2 Procurement of tower

8.2.1 Scope of tender

The scope of tender for procurement of tower covers:

a) Scope of supply.

b) Technical data and services to be provided by the tower manufacturer.

8.2.2 Main technical data

The technical documents, fabrication drawings and acceptance criteria of the tower shall be provided by the wind turbine manufacturer.

8.2.3 Main technical requirements

The tender documents shall specify the following technical requirements:

a) The scope of supply shall include the tower proper, foundation connectors and tower accessories. The tower accessories include the power socket system, lighting system, earthing conductors for lightning protection, cable racks, cable clamps, and junction boxes.

b) The manufacture supervision and acceptance of the tower should be conducted by the wind turbine manufacturer, and shall be entrusted to a qualified third party when the tower is supplied by the wind turbine manufacturer.

c) The tower shall comply with GB/T 19072.

8.3 Installation of wind turbine

8.3.1 Scope of tender

The scope of tender for installation of wind turbine shall cover:

a) Handling, site transport and storage of the wind turbine proper, tower and its auxiliaries, special tools, and spare parts.

b) Check of the dimensions and positions of embedded parts for wind turbine foundation.

c) Installation and commissioning, including installation of the wind turbine proper, tower and its accessories; installation of the control cabinets and cables; installation, inspection and testing of the wind turbine control system; and commissioning and trial run of the whole wind turbine. In installation, the terminal from the frequency converter cabinet or outlet cabinet at the bottom of the tower may be taken as the interface between the wind turbine and the power collection system.

d) Maintenance of wind turbine equipment during the installation period.

8.3.2 Main technical data

The tender documents shall provide the following technical data:

a) Parameters of the wind turbine, including the overall dimensions and transport and hoisting weight of the nacelle, generator, rotor, blade and each tower section.

b) Conditions of the installation site, including the size and ground load-bearing capacity of the hardstand, parameters of the on-site road, and requirements of site transport and assembly for the hoisting equipment.

c) Delivery schedule of the wind turbine proper, tower and other main equipment, and wind turbine installation schedule.

d) Installation process, inspection, testing, acceptance and safety requirements for the wind turbine.

8.3.3 Main technical requirements

The tender documents shall specify the following technical requirements:

a) The installation of the wind turbine shall comply with GB/T 6067.1, GB/T 19568 and JGJ 276 and the requirements provided by the manufacturer.

b) The selection and configuration of the hoisting equipment shall be proposed according to the wind turbine installation height and weight, rotor diameter, site-specific conditions, and installation requirements provided by the wind turbine manufacturer. The hoisting equipment includes the main hoist, auxiliary hoist, vehicles for site transport, as well as the installation tools, safety apparatuses and test equipment to be furnished by the contractor. In addition to GB/T 6067.1 and JGJ 276, the main hoist shall also meet the requirements for maximum hoisting height and maximum hoisting weight.

    c) Occupational health and safety and environment (HSE).

    d) Submission of documents, including the submission time, quantity and approval procedures of the monthly reports, weekly reports, method statement, special construction measures, quality control plan, HSE plan, and inspection and test report.

9 Foundations for wind turbine and its step-up equipment

9.1 Design scheme

9.1.1 Determine the design class of the foundations, and select the design criteria, flood control criteria, and seismic fortification category.

9.1.2 Carry out the wind turbine foundation design based on the geological investigation findings and wind turbine type, including selection of the bearing stratum and foundation type, and design of the foundation structure, ground treatment, monitoring, corrosion control, crack control and embedded conduits.

9.1.3 Carry out the foundation and embedded conduit design based on the technical data on wind turbine step-up equipment.

9.1.4 Carry out the lightning protection and earthing design according to the geological investigation and the requirements provided by the wind turbine manufacturer.

9.2 Scope of tender

The scope of tender shall cover the earthwork, ground treatment, concrete placement and curing, rebar fabrication and installation, installation of embedded parts, cable conduits and drain pipes, internal earthing, corrosion control and sealing and surplus earth transport for the foundations for wind turbine and its step-up equipment.

9.3 Main technical data

The tender documents shall provide the following technical data:

    a) Scale and rank of the project, site natural conditions, basic wind turbine parameters, flood control criteria, and seismic design criteria.

    b) Foundation design scheme, including the requirements for the bearing stratum and for the type, size and material properties of the foundation.

    c) Bill of quantities for the foundation for wind turbine, including the quantities for earthwork, concrete, rebars, embedded parts, cable conduits, earth electrodes, foundation piles, ground treatment, corrosion control, and monitoring.

    d) Bill of quantities for the foundation for wind turbine step-up equipment, including the quantities of earthwork, concrete, rebars, masonry, mortar plastering, embedded parts, cable conduits, and earth electrodes.

    e) Drawings of the foundations, including the general layout, excavation drawing, plan, typical profile, earthing diagram, and embedded piping diagram.

9.4 Main technical requirements

The tender documents shall specify the following technical requirements:

- a) Foundation excavation and backfill, and ground treatment.
- b) Fabrication and installation of embedded parts and rebars.
- c) Fabrication, transport, construction and testing of foundation piles.
- d) Mixing, transport, placement, temperature control and curing of foundation concrete.
- e) Installation of foundation connectors, monitoring equipment, cable conduits, earth electrodes and other embedded parts.
- f) Masonry of the foundation for wind turbine step-up equipment.
- g) Seasonal construction.
- h) Foundation corrosion control.
- i) Earthing resistance of wind turbine earthing mat.
- j) Acceptance of the works.

10 Procurement and installation of wind turbine step-up equipment

10.1 Procurement of wind turbine step-up equipment

10.1.1 Design scheme

10.1.1.1 Determine the connection mode between the wind turbine and its step-up equipment, as well as the capacity and voltage level of the step-up transformer.

10.1.1.2 The influence of environmental conditions shall be fully considered to determine the type of the wind turbine step-up equipment.

10.1.2 Scope of tender

The scope of tender for procurement of wind turbine step-up equipment shall cover:

- a) Scope of supply.
- b) Technical data and services to be provided by the manufacturer.

10.1.3 Main technical data

The tender documents shall provide the following technical data:

- a) Environmental conditions, such as altitude, wind speed, temperature, humidity, thunderstorm, salt mist, hail, icing, snow, solar radiation, and earthquakes.
- b) Model, specifications and quantity of the transformers, HV load switches or circuit breakers, HV fuses, lightning arresters, lighting and maintenance transformers, LV frame circuit breakers, surge protectors, miniature circuit breakers, etc.
- c) Drawings, including the electrical wiring diagram of wind turbine step-up equipment.

10.1.4 Main technical requirements

The tender documents shall specify the following technical requirements:

- a) Material of the enclosure for wind turbine step-up equipment, and IP class of the enclosure and oil tank.
- b) Electrical parameters and mechanical properties of the step-up transformers, HV load switch-fuse combination or circuit breakers, lightning arresters, lighting and maintenance

transformers, LV circuit breakers, surge protectors, miniature circuit breakers, etc.

c) Protection, measuring and monitoring for wind turbine step-up equipment.

d) Types of HV and LV outlets.

10.2 Installation of wind turbine step-up equipment

10.2.1 Scope of tender

The scope of tender for installation of wind turbine step-up equipment shall cover:

a) Handling, storage, installation, testing and acceptance of the equipment.

b) Laying of cables or conductors for the HV and LV sides of the equipment.

c) Fabrication, installation and testing of HV and LV cable heads in the equipment.

10.2.2 Main technical data

The tender documents shall provide the following technical data:

a) Quantity, type, size and weight of the wind turbine step-up equipment, and parameters of its major components.

b) Interfaces of the wind turbine step-up equipment with the installation of wind turbine and with the power collection system.

10.2.3 Main technical requirements

The tender documents shall specify the following technical requirements:

a) The relevant standards and specifications that the installation of the wind turbine step-up equipment shall comply with.

b) The installation schedule of the wind turbine step-up equipment shall be coordinated with that of the wind turbine proper, tower and power collection system.

11 Power collection system

11.1 Design scheme

11.1.1 Determine the voltage level, type, grouping and routing of the power collection system based on the scale of the wind farm, layout of wind turbines, on-site road, topography, and climate.

11.1.2 The power collection system of the wind farm may use cables, overhead lines, or their combination. For cables, direct burial should be adopted, the cables should be laid along the road for a wind farm in a mountainous area, and the type, cross section and length of the cables shall be determined. For overhead lines, the type of the conductors, poles, pylons and foundations shall be determined according to the meteorology, topography, geology, and transport conditions, etc.

11.2 Scope of tender

The scope of tender for procurement and installation of power collection system shall cover:

a) Procurement of cables and cable accessories.

b) Excavation, cable laying, testing, backfill and acceptance of cable trenches for the power collection system that uses buried cables.

c) Fabrication, installation and testing of cable accessories.

d) Construction of foundation for the power collection system that uses overhead lines.

e) Procurement, installation, testing and acceptance of materials for the power collection system that uses overhead lines.

11.3 Main technical data

The tender documents shall provide the following technical data:

a) Environmental conditions, including the maximum and minimum ambient temperatures, annual average temperature, relative humidity, altitude, maximum wind speed, seismic intensity, type and grade of pollution, number of thunderstorm days, and thickness of ice.

b) Topographic and geological data of the wind farm.

c) Type and length of power cables and communication cables, and quantity of cable distribution boxes and cable accessories.

d) Length of overhead lines; specifications, type and weight of conductors and earthing wires; type and quantity of poles and pylons; specification and weight of each type of poles and pylons; and specifications and quantity of fittings and accessories, earthing materials, optical cables, lightning arresters, and fuses or disconnectors.

e) Quantities of earthwork, concrete, rebars, marker posts, protective covers, and sand or soft soil, and specifications and quantity of embedded parts.

f) Bill of quantities for the soil and water conservation and environmental protection works.

g) Drawings, including the layout and electrical wiring diagram of power collection system, plan and profile of overhead lines, details of poles and pylons, assembly drawing of insulator strings, earthing diagram, general charts of pylons and of foundations, profile of cable laying, and layout of cable wells.

11.4 Main technical requirements

The tender documents shall specify the following technical requirements:

a) Scope of tender and definition of interface.

b) Type, voltage level, conductor's sectional area, material, and structure.

c) Packaging, transport, storage and testing of the system.

d) Construction techniques and process for the system.

e) Main technical parameters and construction process for optical cables and accessories.

f) Soil and water conservation and environmental protection during construction of the system.

g) HSE.

12 Road and wind turbine hardstand

12.1 Design scheme

12.1.1 Determine the road design parameters through techno-economic comparison according to the transport parameters and scheme for wind turbines and other large and heavy equipment, wind turbine installation scheme, and wind turbine operation and maintenance requirements.

12.1.2 Propose site access alternatives based on the transport conditions around the wind farm, and then determine the site access scheme through techno-economic comparison.

12.1.3 Determine the on-site road scheme for the wind farm through techno-economic comparison according to the topography, geology, transport conditions, road design parameters and wind turbine layout of the wind farm.

12.1.4 Propose a road reinforcement and reconstruction scheme based on the status quo of the internal and external roads of the wind farm.

12.1.5 Determine the elevation and dimensions of the hardstand according to the topography and geology of the specific wind turbine site, installation scheme, and road elevation.

12.1.6 Carry out cut-fill balance for the wind farm and propose a design scheme for the spoil areas.

12.2 Scope of tender

The scope of tender for road and wind turbine hardstand covers:

   a) Subgrades, pavements, road drainage facilities, retaining walls, slope protection and road auxiliary facilities.

   b) Site grading for the hardstand.

   c) Road maintenance during the construction period.

   d) Construction of soil and water conservation and environmental protection works.

12.3 Main technical data

The tender documents shall provide the following technical data:

   a) Boundary point between the public road and the site access road of the wind farm.

   b) Functional requirements for the site access road and the on-site road.

   c) Requirements for clearance and load-bearing capacity of the road for transport of large and heavy equipment.

   d) Design parameters of the site access road and on-site road such as width of subgrade and pavement, minimum turning radius, curve-widening value, and maximum longitudinal slope.

   e) Length of new and reconstructed site access road and on-site road; surface clearing for road construction; bill of quantities for earthwork of the site access road and on-site road; bill of quantities for pavements, retaining walls, intercepting and drainage ditches, protection and safety facilities; bill of quantities for culverts and other drainage facilities; bill of quantities for reinforcement and reconstruction of bridges and culverts; traffic signs; bill of quantities for spoil areas; and bill of quantities for landscaping of the areas along the roads.

   f) Bill of quantities for hardstand, including surface clearing, earthwork, retaining walls, intercepting and drainage ditches, ground and surface course treatment, and landscaping.

   g) Drawings, including general layout of roads, typical structural drawings of subgrade and pavement, typical longitudinal and transverse profiles of road, structural drawings of slope protection, retaining wall, bridge and culvert, and typical drawing of hardstand.

12.4 Main technical requirements

The tender documents shall specify the following technical requirements:

   a) Construction of roads, bridges and culverts.

   b) Design dimensions of the hardstand.

c) Excavation, backfill and ground and surface course treatment of the hardstand.

d) Location, capacity and construction of the spoil area.

e) Soil and water conservation and environmental protection during construction.

f) HSE.

g) Acceptance of the works.

13 Step-up substation

13.1 Procurement of electrical equipment

13.1.1 Design scheme

13.1.1.1 Determine the main electrical connection, electrical equipment layout and control, protection and communication scheme of the step-up substation according to the planned scale of the wind farm, construction sequence, and requirements for grid connection and operation and maintenance.

13.1.1.2 Determine the specifications, technical parameters and quantity of the main electrical equipment according to the environmental conditions, main electrical connection, substation layout, and short-circuit current calculation results.

13.1.2 Scope of tender

The scope of tender for procurement of electrical equipment of the step-up substation covers:

a) Scope of supply.

b) Technical data and services to be provided by the manufacturer.

13.1.3 Main technical data

The tender documents shall provide the following technical data:

a) Environmental conditions, including the altitude, temperature, humidity, grade of pollution, thunderstorm, icing, wind speed, and seismic intensity.

b) List of equipment and materials.

c) Scope of electrical equipment supply, including main equipment, spare parts, and special tools.

d) Drawings, including electrical wiring diagram, system configuration diagram, and equipment layout diagram.

13.1.4 Main technical requirements

The tender documents shall specify the following technical requirements:

a) Main technical standards applicable to the electrical equipment.

b) Specifications for the major components of the electrical equipment.

c) Packaging, transport, storage and testing of the main electrical equipment.

d) Scope of technical services, list of submittals, and timeline.

13.2 Installation of electrical equipment

13.2.1 Scope of tender

The scope of tender for installation of electrical equipment of the step-up substation shall cover:

a) Handling, storage, installation, commissioning and acceptance of the electrical equipment.

b) Laying, installation and testing of power cables, control cables and optical cables.

c) Fabrication, installation and testing of cable accessories.

d) Earthing.

e) Procurement and installation of embedded conduits, earthing materials, fire blocking materials, steels, coppers, etc.

f) Joint commissioning and trial run of the substation.

13.2.2 Main technical data

The tender documents shall provide the following technical data:

a) Project scale, geographical location, etc.

b) Interfaces of the step-up substation with the transmission line and with the power collection system.

c) Model and quantity of such equipment as main transformers, HV distribution equipment, reactive power compensators, HV and LV switchgear, earthing transformers, station-service transformers, diesel generator sets, busbars and distribution boxes.

d) Specifications and quantity of dispatching and communication equipment, gateway metering equipment, and equipment of such systems as control and protection system, AC and DC power supply systems, wind power prediction and control system, video monitoring and security system and wind turbine SCADA system.

e) Type and quantity of power cables, control cables, communication cables and accessories.

f) Type and quantity of installation materials such as embedded conduits, earthing materials, fire blocking materials, steels, and coppers.

g) Drawings, including general electrical layout, main electrical connection diagram and earthing diagram of the step-up substation, electrical layout of each floor of the buildings, and distribution diagram of video monitoring and security system.

13.2.3 Main technical requirements

The tender documents shall specify the following technical requirements:

a) Handling, site transport and storage of electrical equipment.

b) Installation and testing of electrical equipment.

c) Combined testing of electrical equipment.

d) Earthing of electrical equipment. The measured earthing resistance of the step-up substation shall comply with GB 50065.

e) Acceptance, trial run, and handover.

f) HSE.

13.3 Civil works

13.3.1 Design scheme

13.3.1.1 Determine the step-up substation site based on the wind turbine layout, topography, geology, transport and environmental conditions, direction of the outgoing line, and incoming route of the power collection system.

13.3.1.2 Determine the safety class, design service life, seismic fortification intensity and category of the buildings in the substation.

13.3.1.3 Determine the general layout of the substation as well as the layout, functional zoning, foundation type, ground treatment scheme, superstructure type, architectural style and interior and exterior decoration requirements for each building in the substation; propose the floor space, gross floor area and height of each building; and determine the design scheme for site grading.

13.3.1.4 Determine the design scheme for the buildings (structures), equipment foundation and outdoor works.

13.3.1.5 Determine the design scheme of protection against flood, blowing sand, corrosion and for thermal insulation.

13.3.1.6 Determine the water supply and drainage design scheme.

13.3.1.7 Determine the HVAC design scheme.

13.3.1.8 Determine the electrical design scheme for the buildings.

13.3.2 Scope of tender

The scope of tender for civil works covers:

a) Earthwork, slope protection, and intercepting and drainage works for the step-up substation.

b) Civil works and decoration works for the buildings (structures) in the substation, and procurement and installation of the electrical and plumbing equipment for the buildings.

c) Procurement and installation of lightning protection and earthing devices, water supply and drainage facilities and electrical conduits for the buildings (structures).

d) Construction of soil and water conservation and environmental protection works.

13.3.3 Main technical data

The tender documents shall provide the following technical data:

a) Site and environmental conditions, including altitude, temperature, humidity, thunderstorm, rainfall, wind speed, seismic intensity, and depth of frozen soil.

b) Topographic and geological data.

c) General layout of the substation and functions of various parts of the buildings (structures).

d) Layout, structural type, floor area, height, decoration criteria, and lighting, ventilation and thermal insulation scheme of each building (structure).

e) Bill of quantities for civil works, including the quantities for construction power supply and water supply, site grading for the substation, building construction, framework and foundation for transformation and distribution equipment, water supply and drainage works, HVAC works, lighting, cabling and outdoor works, as well as the specifications and quantity of lightning protection and earth electrodes for buildings.

f) Drawings, including site grading drawing, general layout, general water supply and drainage arrangement, general outdoor lighting arrangement, building views (plan, elevation, and profile), general earthing arrangement, and building lightning protection and earthing diagram.

13.3.4 Main technical requirements

The tender documents shall specify the following technical requirements:

- a) Construction techniques of each building (structure) in the step-up substation.
- b) Construction techniques of water supply and drainage works.
- c) Construction techniques of HVAC system.
- d) Construction techniques of building electrical system.
- e) Seasonal construction.
- f) Earthing.
- g) HSE.
- h) Acceptance of the works.

13.4 Fire protection

13.4.1 Design scheme

13.4.1.1 Fire protection design for the step-up substation shall comply with NB 31089.

13.4.1.2 Determine the design scheme for the fire protection system based on the overall layout of the substation.

13.4.1.3 Determine the configuration of automatic fire alarming system, water firefighting system, and fire extinguishers.

13.4.2 Scope of tender

The scope of tender for procurement and installation of fire protection equipment covers:

- a) Procurement and installation of lighting fixtures for fire evacuation.
- b) Procurement, installation and commissioning of automatic fire alarming system.
- c) Procurement and installation of fire blocking materials.
- d) Procurement, installation and commissioning of water firefighting system.

13.4.3 Main technical data

The tender documents shall provide the following technical data:

- a) Environmental conditions.
- b) Interfaces with other lots.
- c) Specifications and quantity of materials for fire water system, equipment and materials for automatic fire alarming system, emergency lighting equipment, fire extinguishers and fire blocking materials.
- d) Drawings, including the layout of the automatic fire alarming system.

13.4.4 Main technical requirements

The tender documents shall specify the following technical requirements:

- a) Installation of water firefighting system.

b) Installation of fire protection facilities.

c) Automatic fire alarming system.

d) Fire protection for cables.

e) HSE.